YOUR KNOWLEDGE HAS VALUE

- We will publish your bachelor's and master's thesis, essays and papers

- Your own eBook and book - sold worldwide in all relevant shops

- Earn money with each sale

Upload your text at www.GRIN.com
and publish for free

Bibliographic information published by the German National Library:

The German National Library lists this publication in the National Bibliography; detailed bibliographic data are available on the Internet at http://dnb.dnb.de .

This book is copyright material and must not be copied, reproduced, transferred, distributed, leased, licensed or publicly performed or used in any way except as specifically permitted in writing by the publishers, as allowed under the terms and conditions under which it was purchased or as strictly permitted by applicable copyright law. Any unauthorized distribution or use of this text may be a direct infringement of the author s and publisher s rights and those responsible may be liable in law accordingly.

Imprint:

Copyright © 2015 GRIN Verlag, Open Publishing GmbH
Print and binding: Books on Demand GmbH, Norderstedt Germany
ISBN: 978-3-668-07586-3

This book at GRIN:

http://www.grin.com/en/e-book/309179/dinosaurian-the-evolution-of-birds-and-the-origin-of-flight

Karl Stiger

Dinosaurian, the Evolution of Birds, and the Origin of Flight

A Physical Description of Cretaceous Feathered Dinosaurs

GRIN Publishing

GRIN - Your knowledge has value

Since its foundation in 1998, GRIN has specialized in publishing academic texts by students, college teachers and other academics as e-book and printed book. The website www.grin.com is an ideal platform for presenting term papers, final papers, scientific essays, dissertations and specialist books.

Visit us on the internet:

http://www.grin.com/

http://www.facebook.com/grincom

http://www.twitter.com/grin_com

DINOSAURS, THE EVOLUTION OF BIRDS, AND THE ORIGIN OF FLIGHT

A Physical Description of Cretaceous Feathered Dinosaurs

BY

KARL J. STIGER

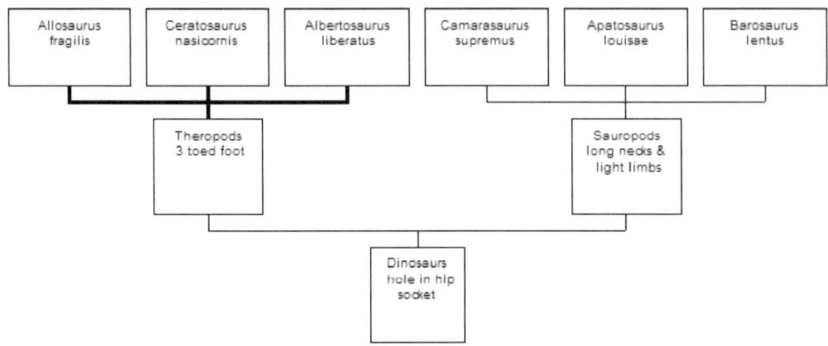

Figure 1.1 Phylogenetic tree representing related groups of dinosaurs within the Saurischian order, specific traits of particular genera are not noted here. However, there are features present in the diagram that do identify the major clades of plant eating & meat eating dinosaurs. Authors own figure.

Abstract

The animals dominated the earth for over 150 million years are called dinosaurs. In them, you can see forms of every major group that ever lived, including birds. In a phylogenetic chart that further describes the evolutionary relations between organisms(e.g., dinosaurs), it's apparent to identify these animals in detail, step by step using physical features, that are evident in the anatomy of the closely related groups that exist among the Dinosauria. Those features that define larger groups are called primitive because they are thought to have evolved earlier, whereas features shared by smaller groups are called derived or advanced. Used in this way, the terms primitive and advanced are relative and do not imply that one feature is better than another. In our chart and cladogram, the character "hole in the hip socket" defines the group dinosaurian, which contains all dinosaurs including birds. Within all vertebrates, having a hole in the hip socket is unique to the dinosaurian, and we say its derived with respect to all vertebrates. Cladistic analysis indicates that the fossil record is probably not complete and that an animal very similar to *Deinonychus* gave rise to birds and then later to dromaeosaurs, including *Deinonychus* and *Velociraptor*. Analyses show that birds evolved from small carnivorous dinosaurs like *Deinonychus* and *Velociraptor* which belong to a group called dromaeosaurs, which lived in the Cretaceous, between 107 and 72 million years ago. Yet, the oldest known bird Archaeopteryx, lived in the late Jurassic about 140 million years ago(Gaffney, 1995).

Introduction

Dinosaurs, capture the imagination because of the fact that, they're bizarre and wonderful. But, except for birds, they're all extinct. And although dinosaur fossils include bones, teeth, tracks and skin impressions and possibly DNA, this evidence amounts to only a tiny fraction of information about when these animals were alive. Saurischian dinosaurs include Sauropod dinosaurs and Theropod dinosaurs. The grasping hand has fingers that differ in size. The thumb is strong and offset. The second finger is longest, and the other fingers become smaller toward the edge of the hand. In carnivorous dinosaurs, the grasping hand developed a number of adaptations including the capacity of flight. Theropod dinosaurs include all saurischian dinosaurs except sauropod's and their early relatives. The advanced feature of these dinosaurs includes a 3 toed hind foot. In tyrannosaurs for example, 3 toes are large while the 5th and 1st toes are reduced or absent. The most advanced group of theropod dinosaurs includes *Tyrannosaurus rex* and *Albertosaurus liberatus*- the Tyrannosaurs. There are several major differences between tyrannosaurs and *Allosaurus*:

I. In *Allosaurus*, all 3 foot bones are about the diameter, whereas, in tyrannosaurs such as *Daspletosaurus*, the middle foot bone became smaller near the ankle.
II. Reduction of the number of fingers to 2, plus the animal's huge body size.

Dilong paradoxus, a tyrannosaur, has even been found with feathers covering its body providing further support for the feathered origin of tyrannosaurs in China during the early cretaceous period 130 million years ago. The feathers on its body were used ti keep the animal warm at night, however, male *Dilong* probably used their feathers during mating displays or signals to rivals, perhaps by flashing brightly colored plumage on the tail, and arms.

Diagnosis of Dinosaur Anatomy

A. Openings for eyes: orbits

i. Openings for nose: nares

ii. Wolf skull: Synapsid = 1 opening behind each of the eye sockets (dinosaurs have 2 openings)

B. Skull of small tyrannosaur = diapsid: 2 openings behind the eye sockets, and an opening between eye and nostril: antorbital fenestrae and 2 others: the laterotemporal and supratemporal fenestrae

iii. Teeth = upper jaw (maxilla) lower jaw (mandible)

iv. Scapula = shoulder girdle, dorsal = back, Sacral = pelvic, caudal = tail

Feathers: an evolutionary perspective

Feathers are similar in both development and composition to the scales of living reptiles, at least that is what scientists thought then. Today, feathers are thought to have originated from a tubular structure called a rachis, which can be described as a distal structure or shaft of the feather that bears the web, from which hook like barbs attach together to form a feather. Two types exist in birds and carnivorous dinosaurs: There are flight feathers- aerodynamically shaped to catch wind drafts for lift and downy feathers- long feathers used to keep the body insulated during cold weather conditions. In *Microraptor*, the feathers were aerodynamic, and used for flight. While *Deinonychus*, *Yutyrannus*, *Dilong* and *Velociraptor* used downy feathers for insulating the body, as well as scaring off rivals and attracting mates.

Maniraptors: Evolutionary trends in becoming a dinosaur?

Just as humans are both descendants or the first primitive and the first mammal, birds are dinosaurs because they are the descendants of the first dinosaur and the first bird. The closest known evolutionary relatives of birds are small carnivorous dinosaurs such as *Velociraptor*, *Ornitholestes* & *Deinonychus*. The group of advanced Theropod's that includes *Deinonychus* and birds are called maniraptors. Maniraptors ("hand robbers"), have an advanced wrist joint, which is formed by a large pulley shaped wrist bone. Some dinosaurs were warm blooded or endothermic – the birds. And if so, when in their history did this feature appear? Several clues in the structure and chemical makeup of the bones seem to suggest that some dinosaurs were warm-blooded. Until a few years ago, the reconstructions of a small theropod dinosaur called *Sinosauropteryx* were plausible. We know these dinosaurs had feathers, so they could have been any color as far as we know. However, feather colors and shape is influenced by cells could Eumelanosomes: long and narrow eumelanosomes correspond to black and gray feather colors, while short and wide eumelanosomes correspond to brown or reddish brown. White feathers however, lack eumelanosomes while iridescent and glossy patterns correspond to narrow eumelanosomes following in the same direction.

Tyrannosaurs

Albertosaurus liberatus is a tyrannosaur quite similar to *Tyrannosaurus rex* its close relative, in evolutionary terms. Both dinosaurs have two fingered hands, and a large skull. But, they differ in proportions of the leg bones as well as their overall size. Tyrannosaur dinosaurs have Gastralia, which are ribs that cover the belly region. And, there are two types: medial gastralia and lateral gastralia.

> "*Prosauropod and theropod lateral gastralia are slightly curved, rod shaped bones, that taper at both extremities. Each lateral gastralium articulates parallel with the medial gastralium along its craniolateral surface, via a groove on the caudalmedial surface of the lateral gastralium. In prosauropods and small theropods such as coelophysids, troodontids, oviraptorids and dromaeosaurs, the lateral gastralia are generally 1.5 to 2.5 times as long their medial counterparts, having a relatively shorter articular surface with the medial gastralium. In large theropods, especially tyrannosaurids, the gastralia usually fuse*" (Claessens, 2004).

Large meat eating dinosaurs of the cretaceous period(e.g., *Tyrannosaurus*) were probably active predators. It was the scimitar like serrated teeth set in sledgehammer jaws that aided them in their predatory habits. A joint in the lower jaw may have helped in absorbing some of the shock generated by struggling prey. The arms of advanced tyrannosaurs were tiny. Scientists in 1902 once thought *Tyrannosaurus rex* had 3 fingered hands like *Allosaurus*, until specimens were discovered which showed advanced tyrannosaur dinosaurs(e.g., *Daspletosaurus*), only had two. Ancestral tyrannosaur lineages (e.g., the branch that contains *Alioramus remotus*) have a more elongated design, I should only point out that more advanced tyrannosaurs such as *Nanotyrannus*, especially as a juvenile, is still in its earlier stage of ontogeny, having un fused plus not yet ossified skeletal bones in addition to laterally compressed teeth. The morphology of *Alioramus altai* suggests that snout elongation was a straight forward process, in which the region of the antorbital fenestrae was simply stretched forward, a change that influenced only the airway in terms of major tissue structures. The change in shape also affected the palatine, in that its dorsal ramus is positioned anteriorly away from its lacrimal. In short snouted tyrannosaurids(e.g., *Gorgosaurus* & *Tyrannosaurus*), the dorsal ramus is positioned closer to the lacrimal above the jugal. In contrast, the entire skull of derived tyrannosaurids has deepened beyond the juvenile condition, and is indicated by a host of characters including the great depth of the Alveolar region of the maxilla, the keyhole shape of the orbits and other proportions of the skull (Brusatte et, al. 2012).

Saurornithoides mongoliensis

Saurornithoides is a small theropod from Mongolia. Like most species of maniraptors, it is only known from a few specimens including the skull. I recognized the resemblance between *Troodon* and *Saurornithoides*, once I examined the *Saurornithoides* skull closely while exploring the hall of Saurischian Dinosaurs during a visit to American Museum of Natural History on Columbus Day of this year. To me, it looked elongated and smooth with large eye sockets, suggesting the owner had excellent binocular vision and could see its prey (small shrew like mammals), in the dark. Holotype: A weathered skull with articulated mandible and associated partial postcranial skeleton. The outer surfaces of the skull and skeleton are highly weathered and, the postcranial skeleton is fragmentary. The skull is missing the dorsal and lateral walls of the braincase and the occiput dorsal to the foramen magnum and most of the frontals were eroded away. The lower temporal and posttemporal bars are missing on both sides except for the suborbital section of the jugal on the right. The specimen includes a maxillary fragment, quadrate, six vertebrae, and a few bones from the feet. Description of Skull: the anterior end of rostrum is rounded in dorsal view. The premaxillary and maxillary structure is indistinct but can be traced from the other side. The internarial bar is wide and dorsally flattened as in other troodontids, *Shuvuuia deserti*, and some ornithomimids. It overlies the nasals above the posterior end of the nares, and ends opposite the posterior nasal border. The distal end of the ascending process appears to separate the nasals anteriorly as in *Byronosaurus jaffei*, *Zanzibar junior*, and *Velociraptor mongoliensis*(AMNH 6515). The premaxilla attenuates posteriorly, and an lateral expansion on the right side is actually formed by the maxilla(Norell et, al. 2009). Description of fossil:

Saurornithoides mongoliensis, is a troodontid from Asia. Like its more familiar cousin *Velociraptor mongoliensis* (made famous by the *Jurassic Park* series), it too shares a sickle claw on the middle toe of each foot, which were probably used to slit the throats of prey. In comparison with the skulls of other dinosaurs(e.g., *Stegosaurus* with a brain weight of 0.3 oz.), the brain of a troodontid(e.g., *Saurornithoides*, *Troodon*) would have weighed 6× more than the brain of *Stegosaurus*. The orbital sockets are large, suggesting *Saurornithoides* had great depth perception, when hunting at night. The estimated length of the AMNH skull is 7.5 in. = 21 cm. And, as a side note: the body of *Saurornithoides* and its relatives were covered with feathers, especially on the arms, these feathers would have insulated it during cold nights in the Mongolian desert, plus males of the same species may have brightly colored ones to advertise themselves to females, during mating dances or to scare off rivals and larger predators.

Velociraptor mongoliensis

The skull was the first ever discovered. One specimen of *Velociraptor* from Mongolia, was thought by some scientists to be intertwined, in a death struggle with a *Protoceratops*- an early Ceratopsian dinosaur. Description of fossil: The skull of *Velociraptor mogoliensis* is intact: the teeth, nares, eye sockets and braincase can be seen, as well as lower jaw(mandible), facing from the left side. The AMNH specimen numbered 6515, has a skull length of 7 in. = 20 cm. *Velociraptor*, like many of its distant and closest relatives(e.g., *Deinonychus*), was a pack hunter with a sickle claw on the middle toe of each foot. Like all maniraptors, it too had feathers covering its body for identification and insulation.

> *The elongate subnaral foramen is pronounced. The antorbital fossa is long and shallow and occupies more than slightly more than half of the snout length. Only the rostral and rostraldorsal portion of its rim is well developed, while the ventral part of the rim is indistinct. The maxillary fenestrae is much smaller. The interfenestral strut separating it from the internal antorbital fenestrae is wide. The orbit is almost circular, only slightly longer, than is high.*

(Barsbold & Osmolsaka, 1999).

Khaan Mckennai:

This oviraptorid (known by a complete skull, and 3 complete skeletons), was discovered by a team of scientists from AMNH in 2001, during an expedition to Mongolia. In 1993, Scientists discovered something surprising, a skeleton of an *oviraptor philoceratops* cocooned inside an egg. Oviraptorids were an unusual group of animals, having parrot like beaks instead of teeth. An unambiguous diagnosis can be made of the clade regarding oviraptoridae and caenagnathidae based on the following characters: a palatal shelf with ventral tooth like projections and a concave pubic shelf. The diagnostic features of this taxon are metacarpal III, not expanding dorsally and not contacting the distant carpals. The nasal is extensively Pneumatized but not dorsally enlarged to forms distinct crest, differing from the morphologically diverse crests that are present on the skulls of *Citipati osmsaki* , *Oviraptor mogoliensis*, *Nemegtomia barsboldi*, and *Rinchenia mongoliensis*(Norell & Baleoff, 2012).

Ornithomimids

Deinocheirus mirificus & Struthiomimus

During the 1960's and 70's , a series of Polish-Mongolian expeditions collected a number of unusual and important specimens. The set of long arms with large claws, are all that is known of *Deinocheirus*, and their immense size indicates that this was a very large animal. The diagnosis of this animal describes it in detail:

> *scapula long and slender; coracoid large; forelimbs long, slender ; humerus straight, twisted, with a pronounced , triangular deltopectoralis crest ; length of humerus equal to about ¾ that of the scapula . Manus only slightly shorter than humerus, with three equally developed digits terminated in claws*

(Osmolsaka & Roneiwicz, 1969).

The AMNH specimen of *Struthiomimus* is shown in a life like pose, highly reminiscent of the way an ostrich stands, except for the mobile outstretched forelimbs and tail extending.

Ornithomimid examples:

Harpymimus	6 prominent teeth in pre maxilla	1.3 m = 4 ft. long
Gallimimus	beak lacking teeth	4.3 m = 13 ft. long

Note: The toothed beaks in ornithomimids were a condition of earlier genera, somehow representing an primitive feature and, was astonishingly unique to the group. Later, advanced ornithomimids had lost their teeth and their jaws modified into a beak. Living birds actually have a combination of feathers and scales , as can be observed for example in the legs of chickens and pigeons. Feathers are necessary for both insulation and flight are merely an advanced evolutionary innovation found in dinosaurs.

What does it take to fly?

In order to fly, an animal needs a variety of physical modifications . The skeleton must be light weight, and the wings needs to move in a specific fashion. Active flight uses lots of energy, and in living vertebrates, flyers are warm blooded- or endothermic. The most distinctive characteristic of birds is flight. Yet, other vertebrates fly. Flight evolved independently in several groups of vertebrates including birds, pterosaurs and some fish.

Conclusion

Animals with 4 limbs are called tetrapods. Dinosaurs in the saurischian order are technically called tetrapods because they have evolved from an common ancestor with four limbs. Some saurischian dinosaurs developed a bipedal stance, because of the innovation of the hole in the hip socket-this joint allowed dinosaurs and even mammals to walk or run with legs tucked under their bodies. In contrast, lizards walk with a sprawling gait and crocodiles walk with bent knees, which differentiates them from dinosaurs, even though they both shared a common ancestor. In essence, Birds are dinosaurs, because they inherited the hole in the hip socket, in addition to 3 toed foot from their theropod ancestors.

Literature Cited

Barsbold, R. & Osmolsaka, H. (1999) *The Skull of Velociraptor(Theropoda) From The Late Cretaceous of Mongolia-* Acta Paleontologica Polonica 44,2 189-219

Brusatte, S. L. , Carr, T. D. , Norell, M. A.(2012) *The Osteology of Alioramus, A Long Snouted and Gracile Tyrannosaurid (Dinosauria: Theropoda) From The Late Cretaceous of Mongolia* AMNH Bulletin 366

Claessens, L. M. (2004) Dinosaur Gastralia: Origin, Morphology & Function Journal of Vertebrate Paleontology 24(1): 89 -106

Gaffney, E. S. , Dingus, L. W. & Smith, M. K. *Why Cladistics?* Natural History; June 95, Vol. 104, Issue 6 pp. 33-35

Norell, M. A, Makovicky P. J. , Bever, G. S. , Balanoff, A.M. , Clark, J. M., Barsbold, R. & Rowe, T. (2009) *A Review of the Mongolian Cretaceous Dinosaur Saurornithiodes (Troodontidae, Theropoda)* American Museum Novitiates Number 3654, 63 pp.

Norell, M. A. & Balanoff, A. M..(2012) *Osteology of Khaan Mckennai(Oviraptorosauria, Theropoda)* Bulletin of the American Museum of Natural History Number 372, 77 pp.

Olsmosaka, H. & Roneiwicz, E. (1969) *Deioncheiridae, A New Family of Theropod Dinosaurs* Acta Paleontologica Polonica No. 21 pp. 6-19